Dr Lloyd Zimba

Questions without Answers? The African Underdevelopment Story

Dr Lloyd Zimba

Questions without Answers? The African Underdevelopment Story

The Zambian Case

LAP LAMBERT Academic Publishing

Cover image: www.ingimage.com

Publisher:
LAP LAMBERT Academic Publishing
is a trademark of
Dodo Books Indian Ocean Ltd. and OmniScriptum S.R.L publishing group

120 High Road, East Finchley, London, N2 9ED, United Kingdom
Str. Armeneasca 28/1, office 1, Chisinau MD-2012, Republic of Moldova, Europe
Managing Directors: Ieva Konstantinova, Victoria Ursu
info@omniscriptum.com

Printed at: see last page
ISBN: 978-620-8-44001-5

Contents

Abstract

Zambia, a Southern African nation, is a land-linked country that shares borders with Angola, Botswana, the Democratic Republic of the Congo, Malawi, Mozambique, Namibia, Tanzania, and Zimbabwe. The country has a population of 19 million people, accounting for 65% under the age of 25 and 52% under the age of 18 (SNDP, 2021). Zambia is also endowed with abundant natural resources such as minerals, forestry, water, and arable land. However, the country is among the poorest nations in the world, recording a poverty rate of above 60 percent and unemployment levels above 6 percent, mainly affecting youths and women. Challenges of poverty, unemployment, and other underdevelopment factors raise a lot of questions on a country like Zambia, which has a small population with all the resources required for development. The problems of underdevelopment are not unique to Zambia because a number of African countries face similar situation, especially those in the sub-Saharan region. A number of factors can be attributed to the cause of these challenges and these include poor governance, domestic and global economic policies, among others. The colonial rule effect also haunts former colonies despite political independence because in spite of political independence, resources of African countries are still under the control of foreign economies under the guise of investment. Foreign investors control terms of investment, trade, and market for raw materials. The population of Africa is growing, and natural resources are diminishing without any development, which raises concerns on how future African generations will survive.

Key Words: Colonialism, Foreign Direct Investment, Natural Resources, and Underdevelopment

Introduction

Challenges of underdevelopment, poverty, and disease in Africa have passed from generation to generation, with each generation hoping to live in a better environment, but nothing has changed. Models, theories, and policies of development have been formulated and different political ideologies have been adopted in an effort to address these problems, but no solution has been found. Despite political independence, Africa is still economically colonised and exists under Western economic conditions. The problems of Africa's development are multidimensional, emanating from natural resource exploitation, poor governance, lack of domestic development plans, loss of human capital, and social conflict. Over the years, the West has exploited the continent's natural resources to the disadvantage of Africa. Africa's natural resources have been and are still taken out in raw form, boosting manufacturing industries overseas. The prospect of development in the region is daunting without taking ownership of its natural resources.

However, there is hope for Africa to develop, and this can be achieved by maintaining a strong economic partnership with the emerging Asian economies. China and India have been pragmatic in their investments in Africa because they promote the manufacturing sector and facilitate technology transfer, and this is slowly changing the development face of the region. Africa is rich in human and natural resources, and its people deserve to live better and dignified lives. The time for a change in economic policy that promoted the development of the continent is now if posterity has to judge the current generation of Africans with respect.

The Big Questions ????????

Why is Zambia and Africa at large poor when they have all the natural resources for development? Why are African countries highly indebted? Why does the continent record high cases of poverty, unemployment, and disease? Why is the region the most affected by climate change? These questions answer themselves. Natural resources are tools for economic growth and development. Natural resources provide capital for poverty reduction, employment creation, and enhance the quality of people's lives. Africa tops other continents in the reservoir of natural resources. The continent hosts 30 percent of the world's mineral reserves, such as cobalt, diamonds, uranium, and copper; 8 percent of natural gas; and 12 percent of oil reserves. It has 40 percent of the world's gold and 90 percent of chromium and platinum. The region has 65 percent arable land. Africa is also a popular investment destination for multinational corporations, which amounted to $45 billion in 2022. The 2022 investment accounted for 3.5% of global foreign direct investment (UNCTAD. 2022 and UNEP 2022).

The continent is also a big recipient of huge humanitarian assistance, loans, aid, and grants worth billions of dollars, while poverty and underdevelopment are still persistent. In 2022, the continent recorded 50 percent poverty among rural populations and 10 percent in urban areas. The region also tops in malnutrition cases, poor health, and poor infrastructure (Statista, 2022). Who is to blame for the underdevelopment of Africa? Within Africa, the sub-Saharan region is the most affected area. It is shocking that the sub-Saharan region, which houses more resources on the continent, is the most underdeveloped area with a high rate of poverty and many other degrading challenges. The region has 30 percent of the earth's mineral resource reserves. It has favourable weather and arable land for agricultural production (UNEP, 2022). The discourse on the cause of the underdevelopment of Africa seems to go beyond the scope of development experts. This can be seen by the continued underdevelopment despite emerging theories and concepts of development focussed on the challenges of the continent. Who is to blame for this predicament? One thing is clear: from the colonial era to date, the natural resources of Africa have been under foreign control. While

Africa is celebrating its political independence, economic colonialism has been ushered in, in the form of investment and international trade. Foreign investors come to Africa and dictate terms of investment, and governments hesitantly agree without choice. What do Africans control? Africa has been reduced to a spectator, hopelessly watching its resources savaged. Africa's ownership of natural resources is not in the hands of African countries, as can be seen. Africa is just a custodian of resources whose ownership and benefits belong to foreign companies. The governments of Africa are powerless to safeguard its resources, either because those in power are complicities of the exploitation or because they have no authority to safeguard them. It is difficult to score development targets in this situation.

To answer the question as to why Africa is poor, is a difficult assignment because there are multiple reasons. Lack of a manufacturing industry is among the causes of underdevelopment. This is because of the repatriation of raw materials by foreign investors who opt to process goods in their countries, where employment is created along the value chain. This has deprived the countries of capital and, in turn, resulted in borrowing. It is also important to indicate that poor countries are suffering from imposed development policies and models. These actions have enslaved the governments of poor nations, making them manage their affairs on an autopilot system devoid of autonomy. This implies that poor countries have to do what the lender wants. Lack of capital in poor countries has caused donors to dictate what to be done, how to be done, when to be done, where to be done, and who to do it. Poor countries are constricted by a number of challenges, including the difficulties in mobilisation of resources for domestic budgets, and as a result, they rely on foreign budget supplements for implementation of capital projects. How can a nation develop when its budget is dependent on external funding? How can a government implement development programmes with resources that are not available? For instance, Zambia's external financing for 2022 accounted for 29 percent, and in 2023 it was 23.9 percent of the national budget (NAZ, 2023; ADB, 2023).

Africa: A Global Market for Products of Its Natural Resources

It is expected, under normal circumstances, that Africa should have a strong economy and a well-established manufacturing sector owing to the availability of natural resources. The continent is supposed to be industrialised, with the highest global gross domestic product (GDP), a high life expectancy, and the lowest poverty rate in the world. However, the opposite is common. Africa's economy mainly relies on capital from raw materials, which are exported at a lower value. For instance, in 2022, Zambia exported 760,000 metric tonnes of unprocessed copper, worth above 5 billion dollars (Muchiya, 2023). In the same year, Zambia's imports of goods were worth 9.05 billion dollars (Statista, 2023). These figures are not supportive of economic growth and development. The expenditure on imports indicates that the country took the foreign exchange out of the country, making it difficult for the country to save capital for its development.

Almost all manufactured goods in sub-Saharan Africa are imported. Starting from motor vehicles, building materials, furniture, electronics, and electric materials to cleaning materials. Shockingly, most of the imported goods are made out of raw materials that are extracted from resources in Africa. Consequently, the country remains without enough foreign exchange for social and economic development. The natural resources the continent has create an opportunity for the enhancement of the manufacturing sector and social and economic development. The manufacturing sector is essential in any economy as it underpins entrepreneurship, boots trade, and generates foreign exchange needed for development. The manufacturing sector further supports employment creation through the value chain, and this contributes to poverty reduction among citizens. The benefits of Africa's resources are visible in foreign countries where investors come from. These resources have developed infrastructure, created employment and have also contributed to the growth of other indus

Africa's Underdevelopment: Is It an Untreatable Disease?

The annals of African underdevelopment status are a puzzle because, as indicated earlier, the continent has all the resources for development. The demise of African development generates different opinions. It may be thought that colonialism could be the cause of the development challenges. However, other continents such as Asia, the Middle East, and South America, which were once colonised, are better off than Africa. Others have blamed African tradition and culture as backward and the cause of the stifled development of the continent. Again, the Middle East and Asian region have stronger traditions and cultural bonds than Africa, and their development is well ahead of Africa. There are those who point to the African education system imposed by the colonial master as the cause of the retarded development of Africa. This is not convincingly true because, for example, all British colonies received similar education systems, and some continents used that education to their advantage, while in Africa the same education system has had little development impact. To further revoke the education theory, almost all British colonies were subjected to a joint Cambridge examination system. Additionally, the west and east have been sponsoring students from Africa to study in their prestigious universities, but this has not impacted much on the development agenda of Africa.

Social scientists and development experts have provided explanations on the issues bordering on the underdevelopment of poor nations and Africa in particular. The second half of the twentieth century was characterised by classic literature on economic growth and development. Classic theories such as the linear stages of growth model (modernization theory) and patterns of structural change, the international-dependence revolution, and the neoclassic or free-market counter-revolution took centre stage as the way to model a development path for poor nations. Theorists tried to replicate the development success of the developed nations stemming from the economic growth that developed countries passed through after World War II, in which industrialised countries such as the USA applied the Marshall Plan, under which huge amounts of

US financial and technical assistance were offered to war-ravaged European countries to rebuild their economies.

In this quest, the USA and European countries poured huge financial and technical support into developing nations, specifically Africa. However, this approach did not have a meaningful development impact on the nations as it resulted in debt accumulation. There was sharp criticism of this development approach from the Marxist school of thought that propagated the neo-colonial dependence model, which blamed earlier development concepts as tools for perpetuating an unequal international capitalist system of rich country-poor country relationships. Some scholars, such as those who promoted the false paradigm model, blamed the underdevelopment of poor nations on western development strategies, which were said to be incorrect approaches (Moyo, 2009; Hoff, H. & Stiglitz, (2000); Murphy et al. (1989); Krugman (1995); Hausmann et al. (2007). Whichever school of thought in this academic battle is presenting the truth, one known factor is that the story of the underdevelopment of Africa is being inherited from generation to generation.

In this trajectory of African development challenges, the voice of the African scholar seemed to be silent. It looks as if African experts do not have solutions to the development challenges facing them. African voices have not adequately been acknowledged, even though they offer workable solutions. This is coupled with African legislatures and policies that do acknowledge domestic ideas. However, policymakers may be excused because they formulate policies in line with donor demands. It is doubtful if these measures really offer a solution to the development problems of Africa (Moyo, 2009; Kachingwe, 2009). Economic targets such as the Millennium Development Goals (MDGs) and Sustainable Development Goals (SDGs) should be hailed as milestones towards changing the status of poor nations; however, the success of these strategies is difficult to measure, and their attainability is difficult to achieve. How can poor nations universally attain equal scores in education, health, energy, water, technology, agriculture, employment, and infrastructure by a given year when they are currently at different levels of development? Development is a process, and it may be difficult for a country to jump from a primary industry directly to a tertiary industry without supporting factors such as capital and manufacturing industries.

How can sub-Saharan Africa attain the MDG and SDG targets when its resources are taken out? Is it attainable for a region that exports raw materials, lacks a manufacturing industry, and is highly indebted and impoverished with a high youth unemployment rate to meet and achieve development targets and goals? The simple answer is a big no! The workable way for these goals to be achieved is to first deal with the cause of these problems. These problems need to be addressed, and then targets and goals assigned according to the economic development category of a particular country. The lower income countries (LIC), lower-middle-income countries (LMIC), and middle-income countries (MIC) should be categorised differently in regards to goals with different target periods.

It is unfair and unworkable that Guinea-Bissau or Malawi should work towards the same targets and goals as China or India within the same period. Holistic achievements of development benchmarks by poor nations amount to academic work.

Africa's state of development is in a critical condition just like a patient with an untreatable chronic disease. It may be speculated that the developed world intentionally stagnates African development for its own economic benefit. This assertion may be justified because all measures that have been imposed on Africa benefit the developed world. An impression has been created that Africa should rely on the developed world so that it continues to be a perpetual dependent and source of raw materials. The question that is still asked is, 'Why should Africa be poor and the poorest in the world?' The solutions to the underdevelopment problems of Africa formulated by the West have fallen short of a cure. The most agonizing and puzzling factor is that the continent has all the tools and resources for development.

Borrowing Your Own Money: An African Dilemma

In the 1960s, after independence of most African countries, indigenous governments took over control of their resources. However, in the 1990s, under the dictation of the International Monetary Fund (IMF) and the World Bank (WB), African governments were forced to liberalise their economies. This enabled private investors to acquire major public economic sectors such as mining, railways, energy and telecommunications. However, privatisation policy and its structural adjustment programme (SAP) resulted in a number of challenges. Several public workers were sent out of employment because new employers did not continue with old employees either because of incompetence or due to the introduction of technology that replaced human labour. It was also a decision by new employers to scale down labour force. In some instances, some companies were liquidated for non-performance. This generated poverty among employees who lost their jobs (Basu, et al., 2000 & Humphrey, 1995).

Africa was later subjected to implementing weak investment laws that gave investors more advantage to exploit natural resources. New economic policies opened doors for foreign investors to solely own company shares. These policies made prevented the establishment of manufacturing industries in host nations and this development turned the continent into an extraction site. These policy weaknesses caused loss of revenue and capital for Africa, leading to borrowing culture in order to supplement its domestic budgets.

Sub-Saharan Africa is the largest recipient of donor funding from Europe, the US, and Asian countries. However, this does not translate into meaningful development. In 2022, donor aid to Africa amounted to $ 204 billion. During the same period, the continent recorded a public debt of US$1.8 trillion (OECD DAC, 2022 & UNCTAD, 2022). Why should governments borrow and accumulate huge debts when the continent has all the resources needed for its development? Africa is supposed to be a major donor-funding continent in the world. The most interesting feature of this borrowing syndicate is that major donors to the continent are the mother-nations of the multinational corporations that have invested on the continent. One would wonder if donor nations use their resources to fund the continent of Africa through aid and grants. It can be speculated that the funds Africa borrow come from its raw materials that are taken out the content. Multinational companies take raw materials from Africa, process them, export finished products to Africa, make money out of them and later lend the money to African governments. The assertion that donor assistance comes from Africa's capital through natural resources that they repatriate from the continent may not be an accusation. It could be the same capital out of Africa's natural resources that they give African governments in the form of aid, grants, and loans. It is vivid that Africa is borrowing money that comes from its own natural resources and it is justified to state that Africa is borrowing its own money!

Debt-Constricted Africa

International financial institutions have been pouring money into Africa since the 1950s, when it was assumed that Africa needed development injections through foreign capital support. In the early years of independence, African governments used much of the resources from donor agents to help citizens through subsidised goods and services (Kean & Wood, 1992). These measures contributed to debt accumulation because countries spent more on consumption without much savings in the economy.

Other causes of the debt crises could be traced to windfall increases in income, particularly from the Organisation of Petroleum Exporting Countries (OPEC) in the 1970s. Huge surplus capital was accrued by the OPEC in Western banks as a result of the rise in oil prices. The Organisation for Economic Cooperation and Development (OECD) countries embarked on imprudent lending behaviours. The OECD countries used the excess capital to extend loans to developing countries without careful credit analysis, and African countries borrowed money to support their public expenditure programmes on the expectation of higher commodity prices in future. Countries were optimistic about the repayment capacity of their debt. However, there was a problem of imprudent borrowing, where countries borrowed to support projects with or without economic returns. These included the implementation of subsidy policies as mentioned earlier or investment in poorly planned projects that failed (Fole, 2003).

The tendency of countries to borrow to address balance of payments was another bait on the debt trap because later African nations failed to repay. It later turned out that domestic production became a measure for the repayment of debts. When it became obvious that poor nations were overburdened by debt accumulation and could not repay, there were calls for debt cancellation under the auspices of highly indebted countries (HIPC) arrangements. Consequently, donor countries and organisations were compelled to cancel the debt (Harold, 1998). As soon as the debts were cancelled, a number of countries entered into another journey of borrowing, resulting in the accumulation of debt again. It is clear that the problem of underdevelopment in Africa cannot be squarely attributed to a single factor. There are intertwined issues that have birthed this problem.

Factors Hindering Development in Zambia and Africa

The journey of poor development in Africa and its causes is not a hidden story. Zambia and a number of countries in Africa have a grim political, social and economic experience. These factors have culminated into the current state of affairs of poverty, disease, instability and poor infrastructure development among other challenges. The cause of this state of affairs are multidimensional. Natural resource exploitation, a lack of a manufacturing sector, and indebtedness are effects of poor domestic politics that have given leeway to the current development challenges. There has been little or no political will by those in government to drive the agenda for development in Africa.

Political and economic colonialism

Africa has interacted with the West since the 15th century during the voyages of European explorers pioneered by Henry the Navigator, Bartolomeu Dias, and Vasco Da Gama, who sailed around Africa (Law, 1979). These expeditious journeys later attracted land explorers such as James Bruce, René-Auguste Caillie, Mungo Park, Heinrich Barth, Richarch Burton, and David Livingstone. The early exploration introduced trade with the indigenous people but later resulted in the colonisation of Africa and the beginning of natural resource exploitation. The indigenous people became cheap labour for the production of the raw materials, which found their way to Europe. Human resources were forcibly taken as slaves, and together with raw materials, they were transported to Europe. The little development Africa saw was around the suburbs of the colonial master, while African communities were left to live in harsh environments.

The unbearable inhuman conditions in Africa led to resistance by indigenous peoples to oppressive colonial rule, and the agitation for freedom developed in the 1960s. Although it may be held that Africa fought and won independence, the reality could be that the colonial master, after much exploitation of the resources, decided to give up rule to the indigenous. Africans had no means to fight the Europeans, who were in control of security and the military. The bow and arrow could not have withstood the automatic machine guns the colonial masters had. If Africans had obtained independence by defeating colonial masters; all colonial masters could have been driven out of the land, their property seized, or they could have been made equal to the indigenous. However, after so-called independence, the colonial master still remained superior in the land. Maybe Africans missed out because their so-called war for independence was mainly focused on freedoms of speech, movements, and social equality. There was no call for economic independence. The colonial master knew that colonisation had not ended. New African governments could not manage economic affairs, as the capital left by the colonial master was being depleted without much development.

Negative economic impact was imminent, and governments had to look back to the former colonial master, who waited in anticipation of the problems. The former colonial master came in the name of an investor.

Changing economic policies by governments became common in trying to address economic woes. Some countries, such as Zambia, which had introduced nationalisation policies after independence, had to abandon them. In the 1990s, the Brent Wood thinkers can be alleged to have thought of taking over control of the whole economies of developing countries when they suggested that the government was not a good manager of the economy and called for private investment. African governments, which were in dire need of financial help after acquiring debts, bowed down to neoliberal policies for a free market under the authority of the International Monetary Fund (IMF) and the World Bank (WB). Privatisation became a popular policy when foreign investors bought government companies. Major public companies were bought by foreign investors because governments had no capital to run them. In the years to follow, foreign direct investment (FDI) became prominent. However, the investment, which was mainly done by the west economies, used Africa as an extraction site. Investors maintained the processing industry in their own country. This hampered real development since technology transfer was limited and jobs were exported through raw materials that were processed outside Africa.

Loss of Africa's Birth Right

Birth right is the concept of things being due to a person upon or by fact of their birth, or due to the order of their birth. These may include rights to citizenship. based on the place where the person was born or the citizenship of their parents, and inheritance rights to property owned by parents or others (Barniter, 2012). Property is an important aspect of a birth right. Throughout history, property has been a great entitlement to individuals, communities, and society at large. At the community and national levels, natural resources constitute collective or communal property for all the people of that particular area. The benefits of the ownership of natural resources as property are manifest in their positive impact on people. The benefits of a national park within a specific community should include providing jobs to the locals and opportunities for investing in that sector.

The benefits of minerals, just like a national part, should provide jobs, shareholding, and other opportunities to local communities. Total ownership of natural resources entails the local community having a voice and authority in their use and management.

However, Africa has no control over its natural resources. The mining, hospitality, agriculture, and major service sectors are in the hands and under the control of foreign economies. This incapacitates its ability to exercise authority over economic and political space. Since the continent has lost ownership of its natural resources, it has lost economic power and capital. Therefore, it has to turn to strong economies for financial help, and assistance is conditioned by the social and political demands of those who help. Help has to be given if the countries meet human rights standards according to the donors' standards. Aid is only possible if the recipient countries promise to invest capital in social and economic areas that the helper wants, even if such conditions hurt citizens. The IMF, the WB and other major economic institutions have weaponized their financial help for Africa, and this has made it impossible for domestically formulated development agendas to materialize. Zambia, for instance, has several times established a divorce and reconciliation relationship with these institutions. After independence in 1964, Zambia enjoyed good economic relations with these institutions under the United Nations Independence Party (UNIP), and in the 1970s, they became estranged as the country opted for nationalisation due to the poor local economic conditions. In the early 1990s, under the Movement for Multiparty Democracy Party (MMD), the country restored its relationship with these financial institutions, but the same party in power parted company with the institutions in the mid-1990s because of rising costs of living and poverty. The New Dawn government of the United Party for National Development brought back these institutions in 2021 with a lot of hope. The country received a lot of kudos from all these institutions and was promised an economic turnaround. At the time, the country was grappling with a huge debt that was defaulted on. The IMF and WB, along with other multilateral partners, set conditions and parameters under which the country would receive urgently needed aid. More demands resurfaced, and deadlines were pushed forward while the exchange rate of the local currency Kwacha and US Dollar started climbing. In 2021, it was K15 to a dollar, and by early 2023, it had climbed to above K27. This sparked a rising cost of goods and services and high inflation, pushing poverty levels to 60% in the country.

It may not be bad or harsh judgement to indicate that Africa owns nothing out of what is supposed to be its birthright. The continent has lost its birthright, and as a result, it has lost respect because it has to seek help from those who possess its entitlements—its resources.

Poor governance system

The African leadership may also bear some blame for the development woes that the continent is going through. A number of African countries have adopted democracy as a system of governance, although some countries are still under military rule. Despite democratic rule in some countries, leaders exhibit authoritarian elements where the rule of law is not observed. In some countries, leaders have manipulated constitutions and extended their tenure of offices, and the rights and freedoms of citizens have also been infringed as people are not able to exercise their freedoms freely. Those with opposing views have been treated as enemies, and the police have been used as a weapon against those in opposition. Government institutions such as the judiciary and police have been interfered with by those in power, resulting in unfair justice. Gender inequality and discrimination are also common, depriving females of the opportunity to participate in development.

Democracy is important in achieving sustainable development as it provides an environment that respects human rights and fundamental freedoms. Democracy enables citizens to express themselves freely and participate in social and economic activities, and it is the bedrock of development as it enables the government to be accountable. Citizens are also able to demand a fair share in the distribution of national resources. It seems as if Africa is not ready to govern itself because it is difficult to comprehend that the region with all the tools for development should be the poorest on a global scale. In most cases, Africa lacks developmental leadership. The political leadership in a number of African countries strives for patronage and nepotism. Those who run governments are either relatives or political sympathisers without regard to qualification and performance. Tribalism and ethnicity in governance are rife among African governments. Tribe and ethnic inclination become requirements for job opportunities and appointments. This has made it difficult for a number of countries in Africa to make progress towards development because such a situation has bred corruption and abuse of office (Woniowei, 2020).

Weak Constitution

Respect for the rule of law is important in the promotion of national development. However, a number of African governments have weak constitutions, and this is a deliberate move to create loopholes for personal interests. Many countries have been changing constitutions with the change of governments to create laws that suit those in power at that particular time. Weak laws have given excessive powers of governance to those in power resulting in the abuse of authority. Amendments to the Constitution are mainly done to incriminate political opponents and not for development purposes. It has become a practice in Africa that each new government spends huge resources on constitution amendment programmes at the expense of national development (Olasunkanmi, 2018).

Corruption

Patronage and nepotism in the governance system have also resulted in corruption. Familiarity in government institutions makes it difficult for transparency and accountability in the running of the affairs of nations. The fight against corruption becomes even more difficult when a relative of a suspect holds high office. Corrupt practices and weak laws have also allowed foreign investors to dictate investment terms in the countries because local officials compromise in policy implementation. The fight against corruption is mainly targeted at those who were in power in previous governments. It has become a practice for law enforcement agencies to wait until one leaves office for investigations to be instituted. Such actions have been deemed political and usually do not yield justice for the accused or government, as cases flop because the accused had time to cover up the case when he or she was in government.

Poor social and economic development policies

Poor countries are regarded as pilot projects for untested ideologies, policies, and programmes. This has reference to the failed theories, models, and policies that have been imposed on poor nations over the years. Since independence, a number of countries in Africa have been subjected to a number of political and economic ideologies, policies, and programmes. The continent had tested capitalism and socialism ideologies. Tanzania and Zambia, for instance, had adopted socialism after independence, which was being implemented alongside the capitalist system. Africa lacks home-grown political and economic ideologies to facilitate its development agenda. The continent receives aid from both the west and east on different political and economic environments. This would have been an economic advantage and opportunity for Africa, as it could have produced a hybrid development system, but the outcome has been a loss of direction on the development path. It should be indicated that democratic and military leaders have ruled different African countries, and both of these systems have bred corruption, patronage, abuse of power, and polarisation of societies with little development. Those in power have amassed wealth overnight, while others have offshore investments, leaving their citizens impoverished.

Further, a number of African countries ride on unworkable Western economic development plans. Sustainable development goals, Vision 2060, and other regional goals are not implementable and are just academic activities in Africa. These programmes have become conduits for misallocation and misappropriation of resources. There is also a problem of lack of continuity for development plans; when leadership changes, unfinished development programmes that were implemented by the previous government are abandoned. For instance, the Zambian government had embarked on the construction of the Chipata-Mchinji railway line under the United National Independence Party (UNIP), which was meant to connect Zambia, Malawi, and Port Nacala in Mozambique, and this was going to offer the shortest route for Zambia to the coast. However, when the new government of the Movement for Multiparty Democracy (MMD) came to power in 1991, the project was abandoned, presumably out of detestation for the former regime. This trend continued when the Patriotic Front (PF) Party came into power in 2011 without regard that sustainable social and economic development requires continuity.

Development in Africa is also affected by a lack of laws that protect local industries. In order for the country's economy to grow, there is a need for a government and policies to protect and support the various sectors. Entrepreneurs, small-scale industries, and other micro-economic sectors require policies that protect them. Foreign investors cannot compete with small local firms because of capital disparities. Once foreign companies come to Africa, they take control of the market and make it difficult for small industries to grow. Developing countries such as Zambia have attracted multinational companies that have invested in mining, agriculture, chain stores, and other consumer goods. The chain stores, for instance, have stifled the growth of small-scale businesses that deal in groceries and other consumer goods. Additionally, foreign chain stores mainly offer foreign products, which are adopted by local consumers.

The growth of the local industry is important to the development of the country because it provides sustainable revenue for the country. In order to protect local entrepreneurs, governments need to define the kind of businesses foreign and local investors should venture into. It should be indicated that small and medium enterprises (SMEs) are the fuel for economic growth and development and set the foundation for future big companies. SMEs are also important as they contribute to job creation and poverty reduction. Additionally, it should be stated that SMEs lack capital and joint venture agenda. Capital is an important asset in the economic growth and development of a country because it boots local investment and contributes to stable revenue in any country. Joint venture increases capital base for business and this would be able make entrepreneurs grow even without government financial support. However, lack of joint ventures and lack of financial help hampers the growth of small businesses. This situation gives an advantage to foreign investors because they have available capital. Another challenge small businesses face is a lack of skills. A number of entrepreneurs lack knowledge in areas of information technology, marketing, and business management. One needs to understand the market trend, supply, customer demand, competition, and risks before engaging in any business activity.

Poor investment in the agriculture sector

The agriculture sector plays a pivotal role in the economic growth and development of a society. The agricultural sector provides raw materials that are important to the local and global economies. Raw materials are critical to supporting the manufacturing industry. The sector also contributes to the creation of a strong supply chain. The import and export of agricultural goods involve transportation from the point of production to the consumer. This contributes to employment creation, income for citizens, revenue for the country, and innovation in the economy. The sector is also important in the promotion of food security, which is good for the economic stability of the country. A country that is food secure promotes healthy living and productivity among its citizens.

However, a number of poor countries do not invested much in agriculture sector. A number of farmers still use traditional tools for production because they cannot access capital for modern equipment. Additionally, the sector is not profitable for farmers since the market is small. Lack of investment in the sector has resulted in low production and food insecurity, leading to famine in countries.

Lack of patriotism

Patriotism plays an important role in the success of any nation. Citizens should develop love of their nations and be able to contribute to the development of their societies. There is a lack of patriotism among citizens in a number of African countries. Some citizens charged with running the public service mainly look at how they can succeed as individuals and do not care about anyone else. They take government as a creature that exists in silos without understanding that government is the institution created by every citizen. Most independence leaders in Africa, such as Kwame Nkuruma, Julius Nyerere, Daniel Arap Moi, and Dr. Kenneth Kaunda, were patriotic and laid the development foundation of their nations. Cases of corruption, disunity, and persecution were rare. There was cohesiveness among the nationals. Currently, the aspect of patriotism has waned, perpetuating corruption and disunity at the expense of national development.

Economic Competition among African Countries

Africa has the potential to develop its economy and become an industrialised region. However, the regional economic environment among countries is inimical to progress. There is economic competition among and between African nations. The development of Africa, just like the development of any other continent, such as Europe, requires collaboration between different nations. Each country should leverage a particular economic sector while depending on another sector from another country. For instance, Tanzania may concentrate on fruit production, while Malawi should produce more maize. In this arrangement, Tanzania is able to offer a market for maize, and Malawi is able to create a market for fruits. This does not mean that any of these countries should stop producing what the other does but they should continue producing but in different quantities, in order to create a market for each other. This approach is able to boost trade and commerce and save capital within the region for the development of the nations.

The situation in Africa is different, as each and every country produces what is produced in another country in maximum quantities, and as a result, there is no local demand or market for what is produced. This development forces countries to export their commodities outside Africa at a low price. African governments should realise that Africa is an emerging market for a lot of goods and services. That is the reason the western world and Asian countries are exporting their goods and services to Africa and investing on the continent. The more than 1 billion people of Africa are a good and big market for what Africa can produce. Just like China has grown its economy through its own large population for the market of local products and services, Africa can develop its economy in a similar manner.

There is a need for economic and trade collaboration among and between African nations. The nations should create an economic dialogue where they should arrange the economic activities of the continent. Each country should identify what it can produce for the market of another country, and then devise a mechanism for trade.

Stability and Instability: Same Status

Peace is the bedrock of any form of development. Investors feel secure investing their capital in an environment that is peaceful and secure. It can be said that one of the reason some African countries are poor is because of the conflicts. However, peace and conflict seem to have the same results of underdevelopment in poor countries. The countries that have been experiencing conflicts and those that have enjoyed stability face the same fate of underdevelopment. For instance, Mozambique, which experienced civil war since 1975, recorded a poverty rate of 64.2 percent in 2022, while

Zambia, which has been a beacon of peace since its creation, recorded a 60.0 percent poverty rate during the same period (AFDB 2023 and Zamstats 2023). Peace and security facilitated development in the West, but in Africa, they don't have the same results. In some countries, conflicts have affected development due to a lack of accountability because of failed institutions; however, in other countries, stability has been used by those in power to abuse power and mismanage the affairs of their nations at the expense of development.

Lack of Rural Development Strategy

Rural Africa is the area most affected by underdevelopment. This area lacks social and economic amenities. Road, education, and health, as well as housing infrastructure, are poor. However, this area has potential for development. One of the economic activities that can jumpstart rural development is cooperative society. Cooperative societies are important to the economic growth of the country, individuals, and community. Cooperative societies deal in different activities such as agriculture, aquaculture, mining, fabrication, and tailoring have economic value for the economy of the nation. Cooperative societies, if well managed, can contribute to employment creation and poverty reduction through resource distribution. Societies can also attract various people to work in sectors such as production, marketing, and distribution along the supply chain. Employment is important because it provides income to the people, which in turn reduces poverty among them.

In terms of gender participation in economic activities, cooperatives can be used to tap women's abilities and address the problem of women's marginalisation in the economy. Women are in the majority in the agriculture sector in a number of African countries; therefore, bringing them into cooperatives may help empower them economically. Support for cooperatives may also encourage locals to diversify their businesses through the adoption of new technology such as mechanised farming and irrigation systems, which may enhance value addition. This may further lead to the establishment of a manufacturing sector in the country.

Since a number of cooperative societies are established in rural areas, they can contribute to a reduction in rural-urban labour mobility as people are able to find employment within rural areas. Developed cooperatives can also attract other economic and social activities, such as manufacturing industries, and they may also attract better education and health infrastructure. They may also contribute to domestic capital for national development and facilitate nation-wide development. However, cooperative societies face numerous challenges in their operations in Africa. There is government interference because the entity has been politicized. Those in power have used cooperatives for political mirage by funding those they can use to their advantage. The other reason for the failure of a cooperative society growth has been a lack of business knowledge among members. A number of cooperatives have been set up merely to benefit from government input support, and therefore they do not have any business agenda. This is coupled with a lack of business knowledge. A cooperative society is a business institution that requires entrepreneurial knowledge in order to succeed. Lack of knowledge makes it difficult for investment. Failure to identify viable business ventures has also been one of the challenges that cooperative societies face. Some cooperatives operate on what others are doing, even if such business is attainable.

Despite the failures, cooperatives can still be revived and become important economic institutions. There is a need to depoliticize the institution and allow its autonomy. This does not mean that the government cannot fund cooperatives. What the government should do is identify cooperatives with attainable business agendas. Since cooperative societies are business ventures, the condition for funding should be convincingly based on the business plan. The cooperative society should be interrogated in regard to what it wants to achieve, how it can be achieved, and the benefits of its achievements. The funding authority should also be enlightened on the medium- and long-term objectives of the cooperative society before funding it. The funding of the cooperative should be based on entrepreneurial objectives and not party inclination. The funding system should be an investment that a recipient cooperative should pay back. A loan is more important than handouts because it encourages recipients to put in more effort to pay it back.

The other solution is to promote business management skills among the members of a cooperative society, such as in the case of small-scale businesses. A cooperative society is a profit-making venture therefore, it is important that those who run it have knowledge. Cooperative societies operate in a competitive business market that needs business management knowledge.in order to stand out. Therefore, those in the management of the cooperatives should have an understanding of market dynamics, and this calls for knowledge and skills in business.

Government involvement in the running of cooperatives should be in the area of guidance and information dissemination as opposed to politics. The ministry in charge should be critically involved in guiding cooperatives and providing information on matters that may affect the sector. Information on climate change, business diversification, value addition, access to markets, competition, and the export of goods and materials needs to be provided to cooperative management.

Collaboration among cooperatives is also an important development for the success of the sector. In order to grow, cooperative societies need to come together and put their resources together. This arrangement enables the harnessing of different skills, which may help cooperatives grow and develop into bigger businesses.

Loss of Human Capital

In order to achieve development, a country needs to invest in its citizens. Human resources that are knowledgeable understand how to drive the development agenda. A number of countries in Africa have implemented universal education development goals so that citizens access education without discrimination, and this has resulted in increased human capital. Human capital is the knowledge, skills, knowledge, good health, and education that someone possesses. In addition to the repatriation of natural resources, Africa is also losing its skilled human resources. Many skilled Africans migrate to the USA and Europe in search of greener pastures. A number of medical personnel, engineers, and other professionals from Africa migrate to Europe and the USA, depriving Africa of skilled practitioners. This rules out an assumption that African education is inimical to development because these migrants to the West drive development agendas in those nations.

Africa is also losing talented citizens in areas of sports, art, and film acting. Great sports personalities in football, athletics, boxing, and art are adopted by the West and change their citizenship. These people have made a huge economic contribution to the countries that have adopted them. This is a needed capital for Africa. Africa remains a foundation and breeding ground for Western economic benefits in many sectors, be the natural or human resources. These are the people who are supposed to drive the development of Africa; however, their migration is a loss to Africa because their skills and knowledge are not utilised on the continent. (Deming, 2022).

Who controls Africa's resources?

Political independence has not translated much into the agenda for social and economic development. Economic empowerment is important if citizens have to enjoy inherent rights. However, despite freedom from political oppression, most African countries are economically under foreign influence. The International Monetary Fund (IMF), the World Bank (WB), the European Union, and many other western institutions dictate the economic direction of poor countries. This can be seen in terms of trade, commerce, control of the local currency, and the determination of the market for goods from poor countries. It is not surprising that England, which is not a copper-producing country, hosts the London Metal Exchange (LME), where copper prices and markets are determined. It is also not surprising that coffee, which is mainly produced in East Africa, is processed in Europe. Africa lacks ownership of its own resources because even the policies regarding the management of these resources are formulated by the developed nations.

African leaders should change their attitude if the continent has to develop. The continent should have ownership of its natural resources if it has to get benefits. Why should a foreign investor own 100 percent of a local company? Foreign companies have taken over the economic territory of most African nations. They dictate how much they should pay employees. Can an African investor dictate the condition of wages in the USA, Europe, or China? The answer is absolutely no! In the developed world, the investment law dictates what an investor should do. In Africa, investors dictate what the investment law should do. Who owns the power between an investor and the law in Africa? It is an investor. Developing countries are rendered weak as they do not have a strong voice for their resources. Foreign investors exploit employees through meagre wages, which cannot sustain anyone. The labour laws do not offer any tangible solution to the conditions of employees. In Zambia, the minimum wage for general workers was about $77 in accordance with Statutory Instrument Number 70 Order of 2018. The country has a number of investors under weak labour laws that increase profits and subject employees to suffering (Editor, 2022). Africa is also faced with challenges of illegal exploitation of natural resources by citizens who collaborate with foreign dealers. Activities such as illegal mining and timber harvesting are common and unaccounted for. All these resources find their way into the hands of foreign dealers and are repatriated out of the country untaxed.

Natural Resources and Industrialization

Natural resources are the bedrock for economic growth and development and play a critical role in the process of industrialization because they enhance the development of the manufacturing sector. Industrialization transforms an economy from primary agriculture to service sectors, in which individual manual labour is often replaced by mechanised mass production and craftsmen are replaced by assembly lines. Industrialization is important because it enhances more efficient division of labour and the use of technological innovation to solve problems, as opposed to dependency on conditions outside of human control. Industrialization helps create employment through the manufacturing sector. The resources of Africa have the capacity to establish a domestic manufacturing sector and help the continent towards industrialization (Rasure, 2021). Most African countries still have primary agricultural economies with low mechanization. The largely peasantry farming is characterised by low production and low foreign exchange from the produce because of a lack of value addition. The agriculture sector is the largest employer in countries like Zambia, but it does not offer enough income to those involved because it is mainly based on peasant and subsistence scales (World Bank, 2021).

Foreign Direct Investment: Its Impact and Development Implications in Africa

Since the change of economic policies in the early 1990s, a number of African countries, such as Zambia, adopted a free market policy, which saw an influx of foreign investment as multinational cooperation took over major companies on the continent. Foreign direct investment (FDI) was viewed as a huge boost to the economic growth and development of many countries. FDI is an ownership stake in a foreign company or project made by an investor, company, or government from another country. This is a business action to acquire a substantial stake in a foreign business or to buy it outright in order to expand operations in a new region (OECD). In 1919, Eli Heckscher developed theories of foreign investments, followed by Bertil Chlin in 1933 using neoclassical economics and macroeconomic theory. However, it was Stephen Hymer who specifically dealt with the theory of FDI in 1960 (Buckley, 2011). They found out that among the reasons for FDI was the need to expand business beyond domestic boundaries, search for new markets, compete in a global environment, and introduce products and goods in other regions.

Importance of FDI

FDI generally promotes the development of the host country. This can be achieved by the transfer of capital and technology from the country of the investor to the host country. Due to the industries that are formed, FDIs also assist the host country's industrialization process. Additionally, FDI generates employment possibilities that generates money for the populace. Due to the increasing output brought on by significant capital and technological investment, FDIs also an aid in economic growth and have a favourable impact on the GDP. FDI is critical to the development of infrastructure, including transportation, communication, educational, medical, and housing facilities, and this plays a significant role in economic development. This results in poverty decline and improvement in people's quality of life. FDI also generates advantageous spill-overs and connections for the nearby area and linkages to the local economy and entrepreneurship (Calimanu, 2021).

Why Has FDI Not Impacted Much in Africa?

FDI offers different opportunities for the host nations to improve their economies. Capital inflows, technology, and skill transfer are important factors in jump-starting national development. Africa is the recipient of millions of dollars through FDI. Zambia is one of the recipients of FDI that in 2022, FDI inflows in to the country reached $116 million, and this is a huge capital investment in the country which can change its economic growth and development status. This is because of the natural resources the country has. Zambia is the third-ranked copper-producing country in the world, and the country is also rich in other minerals, such as emeralds, gold, diamonds, and lithium, in addition to forestry and agriculture resources. Copper accounts for almost 70% of Zambia's exports (UNCTAD, 2023). These resources attract different multinational companies to invest in the country. However, despite these investment inflows, not much development has been seen in Africa.

Zambia Foreign Direct Investment - Historical Data

Year;	*Inflows US*	*% of GDP*	*Year*	*Inflow US$*	*% of GDP*
2021	$-0.82B	3.72%;	2020	$-0.17B	0.95%;
2019	$0.55B	2.35%;	2018	$0.41B	1.55%;
2017	$1.11B	4.28%;	2016	$0.66B	3.16%;
2015	$1.58B	7.45%;	2014	$1.51B	5.56%;
2013	$2.10B	7.49%;	2012	$1.73B	6.79%;
2011	$1.11B	4.73%;	2010	$1.73B	8.53%;
2009	$0.69B	4.53%;	2008	$0.94B	5.24%;
2007	$1.32B	9.42%;			

Macrotrends (2023)

Statistically, this chart shows huge inflows to the country that could have changed the development status of the country. But why has the country remained static in development with such big investments? There are several explanations for the reasons why the inflows have not resulted in meaningful development. It is obvious that the figures of inflow do not translate into liquidity in the local economy, but they are spread across a number of other investment activities, such as consultancy costs, insurance services, and equipment, and what is deposited in local banks is insignificant. This implies that the host country benefits little from these investments, and this is also coupled with weak investment policies that create loopholes for investors to evade taxes and under declare production and profits. The tax holidays and exemption policies offered to investors also rob host nations of capital for development. The economic benefits of the host nations are further undermined by a lack of the manufacturing industry as investors export raw materials to their countries, where they create jobs through value addition.

Import-Dependent Economies

Natural resources provide material for manufacturing, and Africa can boast of all that it has for the manufacturing industry. However, the export of raw materials inhibits the growth of the manufacturing sector in Africa. It can be speculated that this could be a deliberate move by the developed world because it wants to make Africa its market. This situation has resulted in economies in Africa depending on imports. Imports take foreign exchange out of the country. This is a web of economic challenge affecting poor countries because weak economies send their capital to developed countries. In a number of sub-Saharan regions, countries are dependent on imports. Everything from electronics, clothing, and shoe brushes to ear buds and blooms is imported. Almost every product that is found in Africa is imported and traces its raw material from Africa. This works against development, and it is difficult for Africa to come out of its development backwardness.

A Victim Without Reparation

The history of Africa is evident from one generation to another. During 15th century Africans were forcibly taken away from their land and relatives and shipped in inhumane water vessels parked like raw materials to Europe and the USA. In those countries, they were treated like sub-humans and this was followed by the colonial era, a replica of the slave trade. The colonial master dehumanised Africans in their own land without regard to their rights, and this began another episode of natural resource exploitation and repatriation. All these serious crimes against humanity on Africans have not received any United Nations attention for reparation, and yet the UN claims to stand for the promotion of human rights. Germany was made to compensate France and other countries for its invasion of these countries during the first and second World Wars. There have been compensations in other continents on minor cases less serious than the heinous crimes committed against Africans. However, instead of reparation obligations, Africa is being made to pay debts to the West for the loans and aid that have its source from the continent's natural resources.

Africa has the capacity to develop and industrialise with its own natural resources because it has all the natural and human resources at hand. However, this is not possible in this global order where the continent has been turned into an extraction site for the West economic powerhouse. Therefore, this calls for change of African thinking. Africa should make a choice for its future. A choice, either to continue on this current status of being Western-dependent and continue to be a victim of exploitation or liberate itself by changing its economic path and partnership. The continent does not need to ask for a fair share of its resources but claim the benefits of its resource fortunes. It is therefore, envisaged that this is only possible if the continent consolidates its economic ties with Asian countries such as China and India.

The West and Africa: Predator and Prey Relationship

It seems the long relationship between Africa and the West is a prey-predator relationship where the West relies on Africa for resources and Africa relies on the west for aid, while the West gets more benefits. If there is any region that has benefited more from African resources, it is the West, and this has been done at the expense of African development. Today the continent remains the most impoverished of all the five continents. It is not an allegation to indicate that to a large extent the development of the West, points to African resources. It is also not a bad judgement to state that the development of the West has caused the underdevelopment of Africa.

However, there are still prospects for Africa to develop. China, India, and other Middle East nations have shown signs of hope for Africa. Since China opened its economy to the world in 1978, a number of Chinese investors have established their economic investments in Africa. India has also massively invested on the continent. These Asian economic giants have immensely contributed to the developing agenda of Africa in areas of transport, health and education infrastructure, technology transfer, and the manufacturing industry. Asian investors have locally established manufacturing industries in clothing, metal processing, motor vehicle assembly, and building materials which have resulted in revenue and foreign exchange for African governments, employment creation, poverty reduction, and the promotion of the industrialization process. These investments have also created an international market for various goods that are manufactured in Africa.

The Asian-Africa relationship has been premised on shared values, common wealth, and co-existence, and it is not dictated by the inimical conditionality reminiscent of the West. The West-Africa relationship is always characterised by sanctions, withdrawal of aid, and attaching humanitarian help to certain conditions on the presumed nations working against certain principles, even if those principles are deemed unfavourable to the begging nation.

Focus on the East: China, India, and the Future of Africa

African governments should enhance their sustainable development agenda by jealously guarding natural resources against wasteful repatriation and depletion. The present generation should use resources wisely, realising that future generations' survival depends on the same resources.

What's Next for Africa When Natural Resources Deplete?

There will be a time when these resources will deplete, and no one knows what Africa will become. If the region cannot make use of the resources at hand to develop the region, what will it use when the resources are finished? It can be speculated that without natural resources, Africa will be poorer than today. African leaders have been economically influenced by developed countries. Annually, African leaders are invited to economic forums by different developed nations. The leaders of Africa are regular attendees at American, European, and Russian economic forums. These economic forums are mainly focused on Africa's resources. There isn't much that Africa benefits from these gatherings. African leaders attend the meetings to solicit for help while these developed nations look for investment opportunities that would generate capital which should be given to Africa in the form of loans, grants, and aid. Undoubtedly, Africa is a much-sought-for commodity by the giant economies for their economic interests (Roscongress, 2023).

Realigning Economic Relationships

Africa does not need to sever its economic relationship with the West but lobby for mutual benefits. However, maximising opportunities for relationships with Asia has proved to be beneficial to Africa's development agenda. It is evident that the current Western economic and investments policies on Africa serve the interest of the West. The win-win social and economic policies by China and Indian and Middle East investment strategies in Africa are promoting technology transfer as well as manufacturing sector which are crucial to the industrialization of the continent.

Conclusion

Africa deserves development, and Africans deserve dignified lives because of the abundant resources the continent has. However, the current uncontrollable repatriation of resources out of Africa has put the prospects of the continent's development in a quagmire. It is not yet known how Africa's development will be in future. However, there is hope for a prosperous African future. Governments in Africa should change their governance approach with regard to natural resource management. Governments should strengthen its institutions to prevent corruption which has been a gateway to resource exploitation and mismanagement. There is also need for a strong governance system which would promote accountability for those who hold positions of responsibilities. African governments should implement economic and investment policies that are aimed at protecting natural resources for the benefit of citizens. It is time for exploitation of resources to be stopped and Africa took ownership of its natural resources in order to develop a manufacturing sector that would lead to the industrialization of the continent and create a better place for all generations. This is achievable if Africa builds a strong economic development alliance with Asian emerging economies such as China and India who have already demonstrated their willingness to contribute to the development of the region.

Reference

ADB. (2023). Tanzania Economic Outlook. Accessed on 7-11-2023 from https://www.afdb.org/en/countries-east-africa-tanzania/tanzania-economic outlook.

AFDB. (2023). Mozambique Economic Outlook. Recent Macroeconomics and Financial Developments. Accessed from https://www.afdb.org/en/countries/southern-africa/mozambique/ economic –outlook.

Barniter, B. (2012). The Politics of the (lm) Possible. Utopia and Dystopia Reconsidered. P.69).

Basu, A., Calamitsis, E.A. & Ghura, D. (2000). Promoting Growth in Sub-Saharan Africa Learning What Works. IMF. Accessed from https://www.imf.org/external/pubs/ft/issues/issues23/index.htm.

Brantlinger, P. (1985). Victorians and Africans: "The Genealogy of the Myth of the Dark continent". Critical Inquiry. 12(1): 166-203.doi:10.1086/448326.

Buckley, P. J. (2011). The Theory of International Business pre-Hymer. Journal of World Business. 46(1): 61-73. Doc:10.1016/j.jwb.2010.05.018.

Calimanu, S. (2021). Benefits And Advantages of Foreign Direct Investment. https://researchfdi.com/resources/articles/benefits-fdi-foreign-direct-investment/.

Deming, D. J. (2022). Four facts About Human Capital. Journal of Economic Perspectives. 36(3).pp. 75-102.doi:10.1257/jep.36.3.75.

FOCAC. (2023). The 16th Senior Officials Meeting of Forum on China-Africa Cooperation Held in Beijing. Accessed on 10-11-2023 from https://www.fmprc.gov.cn.en/wjb_663304/zygy_663314/gyhd_663338/202310/t20231024_11167121.html.

Fole, A. G. (2003). The Historical Origin of African Debt Crisis. Eastern Africa Social Science Research Review. Michigan State University Press vol.19 (1).pp.59-89.

Harold,J. J. (1998). Developing Countries: Status of the Heavily Indebted Poor Countries Hipc Debt Relief Initiative. DIANE Publishing.pp.2-10.

Hausmann, R., Rodrik, D., & Velasco, D. (2007). "Growth Diagnostic," in One Economics, Many Recipes: Globalisation, Institutions and Economic Growth, by Dani Rodrik. Princeton. NJ: Princeton University Press.

Hoff, H. & Stiglitz, J. E. (2000). "Modern economic theory and development," eds. Gerald M. Meier and Joseph E. Stiglitz. New York. Oxford University Press.

Horn, J., Rosenband, L. & Smith, M. (2010). Reconceptualising the Industrial Revolution. Cambridge MA. MIT Press. London. UK.

Humphrey, P.B. M. (1995). The Limits of Privatization in Africa: Some Preliminary Observations. Journal for Public and Nonprofit Services. pp.187.200. Accessed from https://www.jstor.org.stable/20763798.

Kachingwe, K. (2009). Zambia: Marsha Talks About Millennium Development Goals. The Times of Zambia. Ndola. Accessed on 14.11.2023 from https://allafrica.com/stories/200903091 1.html.

Kean, S.A. & Wood, A.P. (1992). Agricultural Policy reform in Zambia: The dynamics of policy formulation in the Second Republic. Food Policy 17 (1), 65-74.

Krugman, P (1995). Development, Geography, and Economic Theory. Cambridge, Mass. MIT Press.

Law, C.C. (1979). North Africa in the Period of Phoenician and Greek Colonization. In Fage,J.D. (ed). The Cambridge University of Africa.vol.2. Cambridge University Press. p. 135.

Lusaka Times. (2022). Minimum Wage. Accessed from https://www.lusakatimes.com/2022/12/27/ government –increases-the –minimum – wages-for-domestic-workers-shopkeepers-and-general-workers/.

Macrotrends.(2023).https://www.macrotrends.net/countries/ZMB/zambia/foreign-direct-investments.

Moyo, D. (2009). Dead Aid. Why Aid Is Not Working and How There Is A Better Way For Africa. Farrar, Straus And Giroux. New York.

Muchiya, T. (2023). Zambia's 2022 Copper Production Drops. Zambian Business Times. Accessed from https://Zambianbusinesstimes.com/zambias-2022-copper-production-drops.

Murphy. K.M., Shleifer, A., & Vinshny, R.W. (1989). "Industrialisation and the big push," Journal of Political Economy 97 : pp.1003-1026.

NAZ, (2023).2023 Budget Analysis. Accessed from https://www.parliament.gov.zm), Tanzania had a current account deficit of 4.8 per cent of its gross domestic product (GDP).

OECD-DAC. (2022). Official Development Assistance. Accessed from https://www.data.one.org/topics/official-development-assistance.

OECD. (2021). Organization for Economic Co-operation and Development. iLibrary, "Foreign Direct Investment". World Bank. https://www.worldbank.org/en/country/rwanda/overview.

Olasunkanmi, A. (2018). Constitution without constitutionalism: interrogation the Africa experience. Accessed on 13.11.2023. from DOI:10.15406?ahoaj.2018.02.00069.

Rasure, E. (2021). Industrialization.

https://www.investopedia.com/terms/i/industrialization.asp

Roscongress. (2023). Russia-Africa. Second Summit Russia-Africa Economic and Humanitarian Forum For Peace, Security and Development. Accessed on 10-11-2023 from https://summitafrica.ru.en/.

SNDP, (2021). Government of The Republic of Zambia. Lusaka. Zambia.

Statista. (2023). Zambia: Imports of Goods from 2012 to 2022. Accessed from https://www.statista.com/statistics/45772/import-of-goods-to-zambia.

Statista. (2022). Africa: Share of Global Poverty by Country 2023. Accessed on 9-11-23 from https://www.statista.com.

UNCTAD. (2023). FDI To Zambia Reached USD 116 Million in 2022 After Negative Values in 2022. World Investment Report 2023. Accessed on 16-11-2023 from https://www.zambiainvest.com/economy/fdi-inflow-2022-world-investtment-report/.

UNCTAD. (2022). Investment Flows to Africa Dropped to $45 Billion in 2022. Accessed on 9-11-23 from https://unctad.org/news/investment-flows-africa-dropped-45-billion-2022.

UNCTAD. (2022). Public Debt. Prosperity for All. Accessed from https:// www. Unctad.org.

UNEP. (2022). Our Work in Africa. Accessed on 9-11-2023 from https://www.unep.org/regions/africa/our-work-africa.

UNIDO. (2016). The Importance of manufacturing in economic development: Has this changed? Accessed from https://www.unido.org.

Woniowei, F.D.D. (2020). The Nexus between Poor and Bad Governance, and Sub-National Conflicts in Africa. Open Journal of Political Science.vol.10(4).

Zamstats. (2023). At National Level. The incidence of Poverty was estimated at 60.0 percent in 2022 compared to 54.4 percent in 2015. Accessed from https://www.zamstats.gov.zm/highlights-of-the-2022-poverty-assessment.

FSC
www.fsc.org
MIX
Papier | Fördert gute Waldnutzung
FSC® C083411

Kolibri 360 GmbH
Ferdinand-Jühlke-Straße 7
99095 Erfurt, Deutschland
produktsicherheit@kolibri360.de

Druck:
CPI Druckdienstleistungen GmbH
im Auftrag der
Kolibri 360 GmbH
Ferdinand-Jühlke-Str. 7
99095 Erfurt